Early Math Workbook for kids
Counting Numbers
Match , tracing, Write

By Nina. Packer

Copyright: Published in the United States by Nana Packer
Published July 2018

ISBN-13: 978-1722964337

ISBN-10: 1722964332

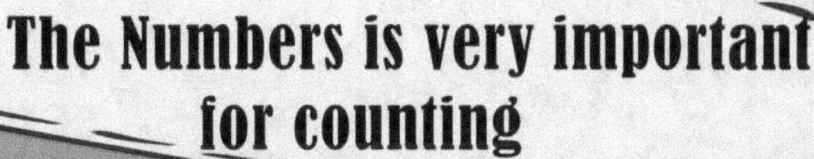

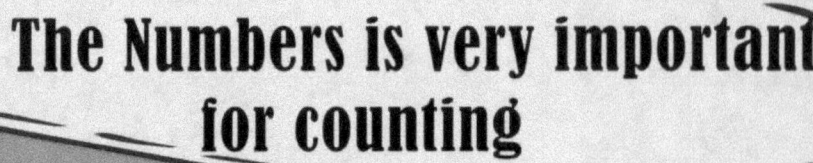

1

ONE

One Butterfly

TWO

Two Rabbits

THREE

Three Elephants

FOUR

Four Seahorses

5 FIVE

Flamingos

SIX

six cows

7

SEVEN

Seven Monsters

8

EIGHT

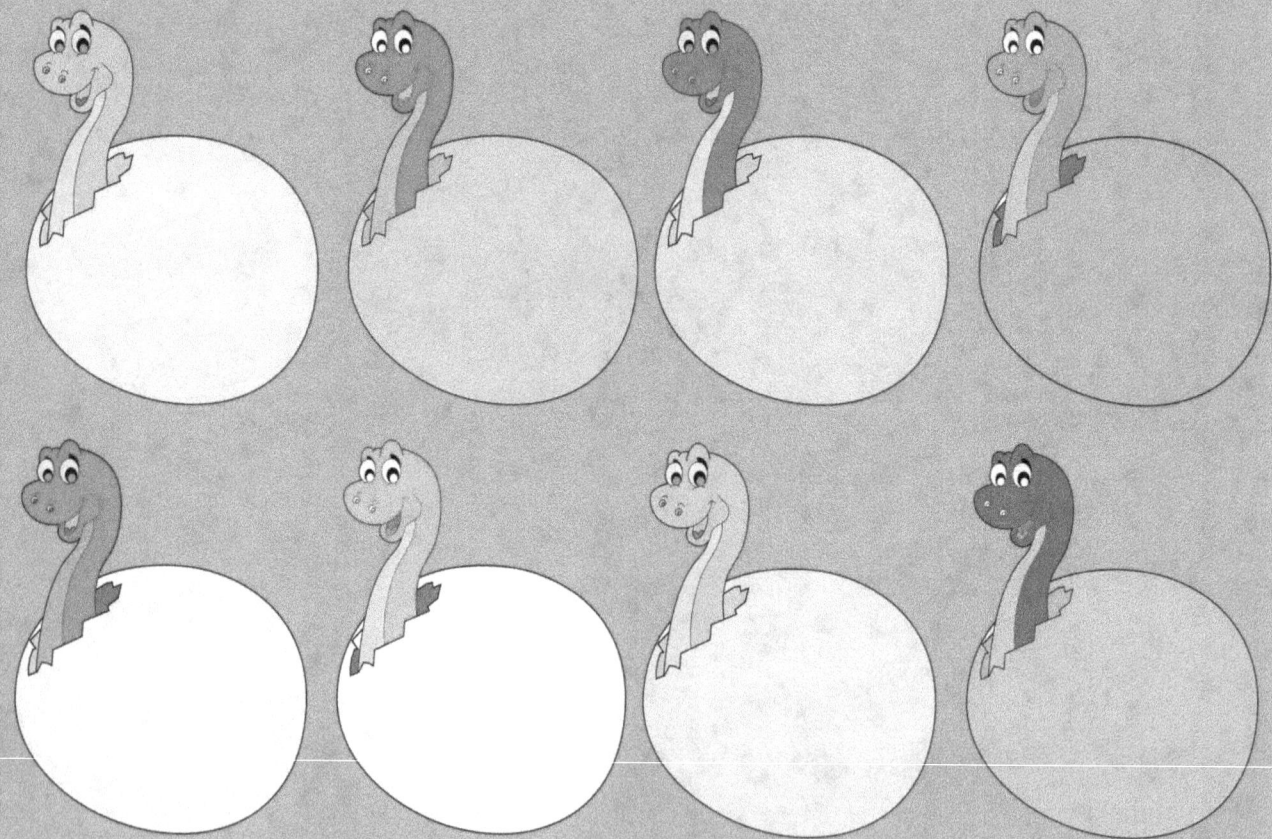

Eight Baby Dinosaurs

NINE

Nine Lions

10 TEN

Ten Octopuses

Name

ZERO

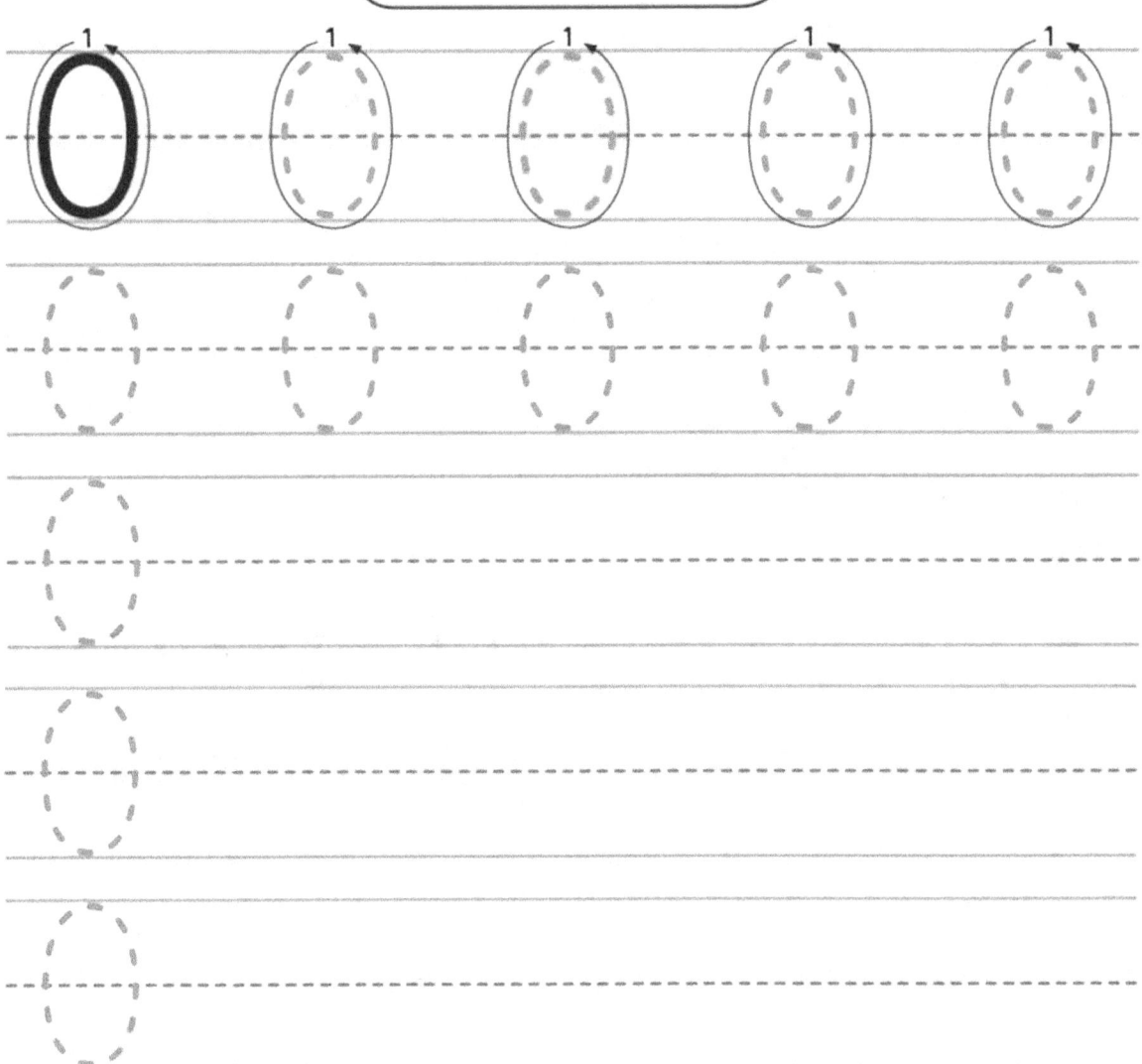

Name

1↓   1↓   1↓   1↓   1↓   1↓

Name _____

2 2 2 2 2 2

Name _____

THREE

3 3 3 3 3 3

Name

4

Name

FIVE

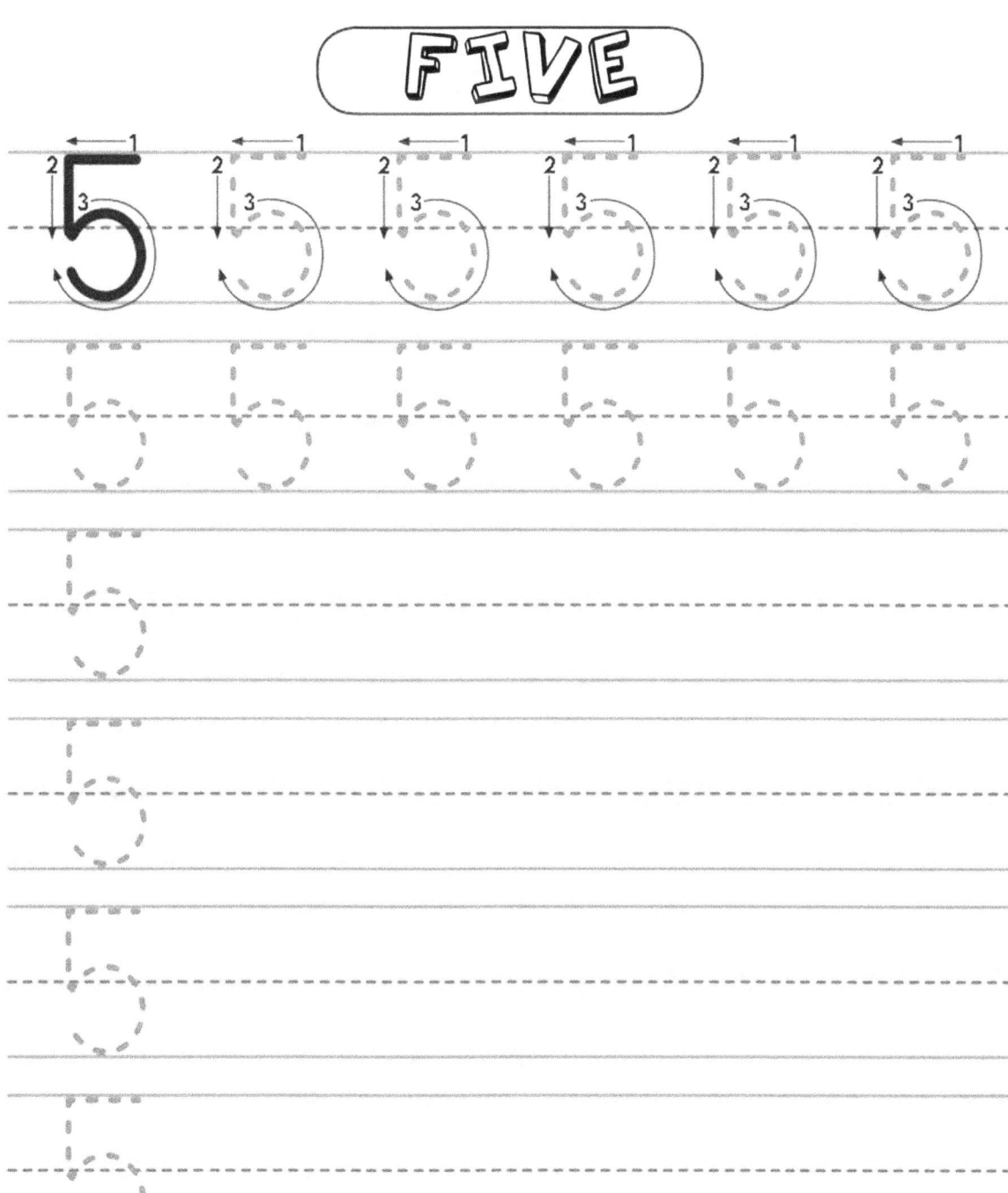

Name

SIX

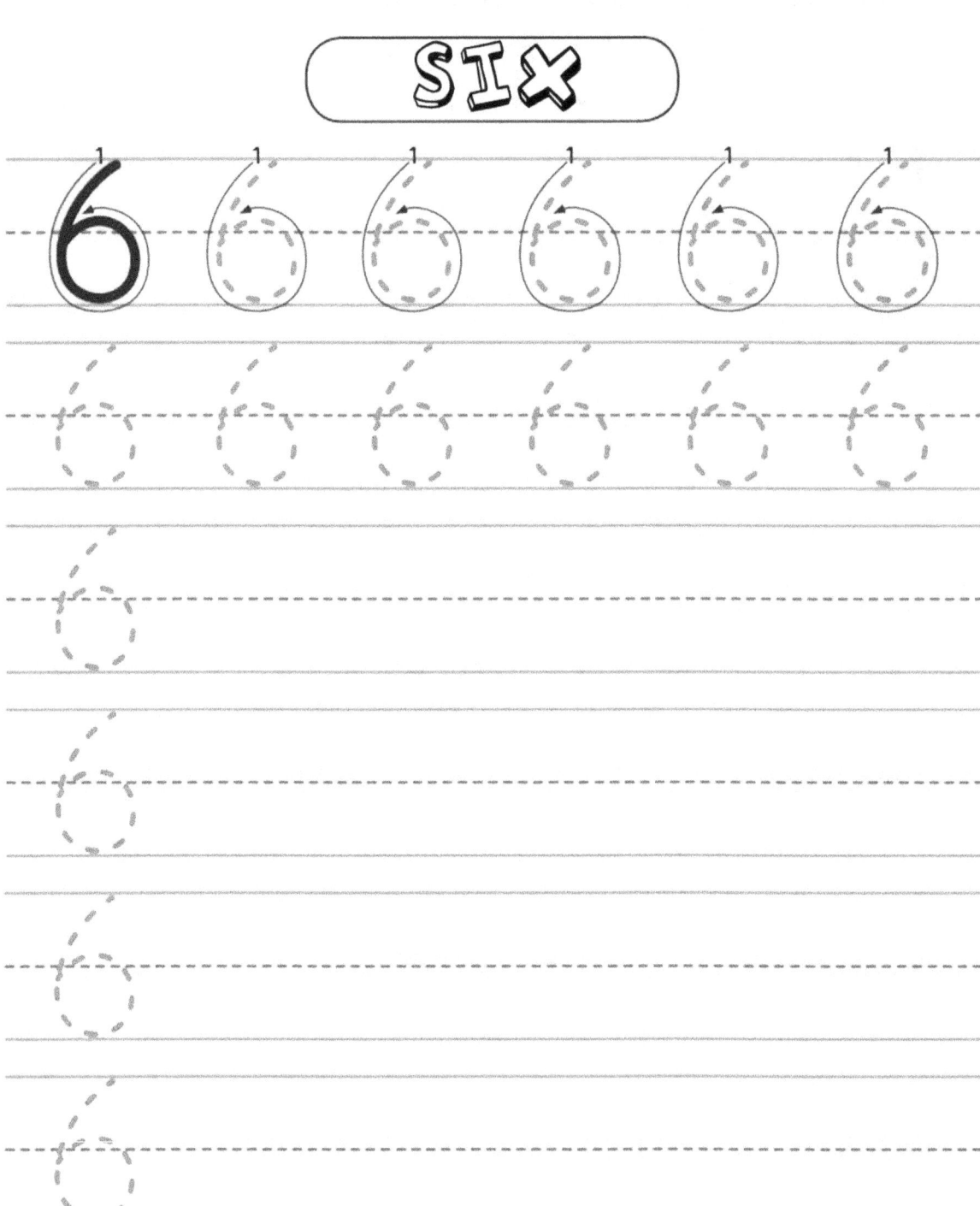

Name

7

Name _____

EIGHT

8 8 8 8 8 8

Name

NINE

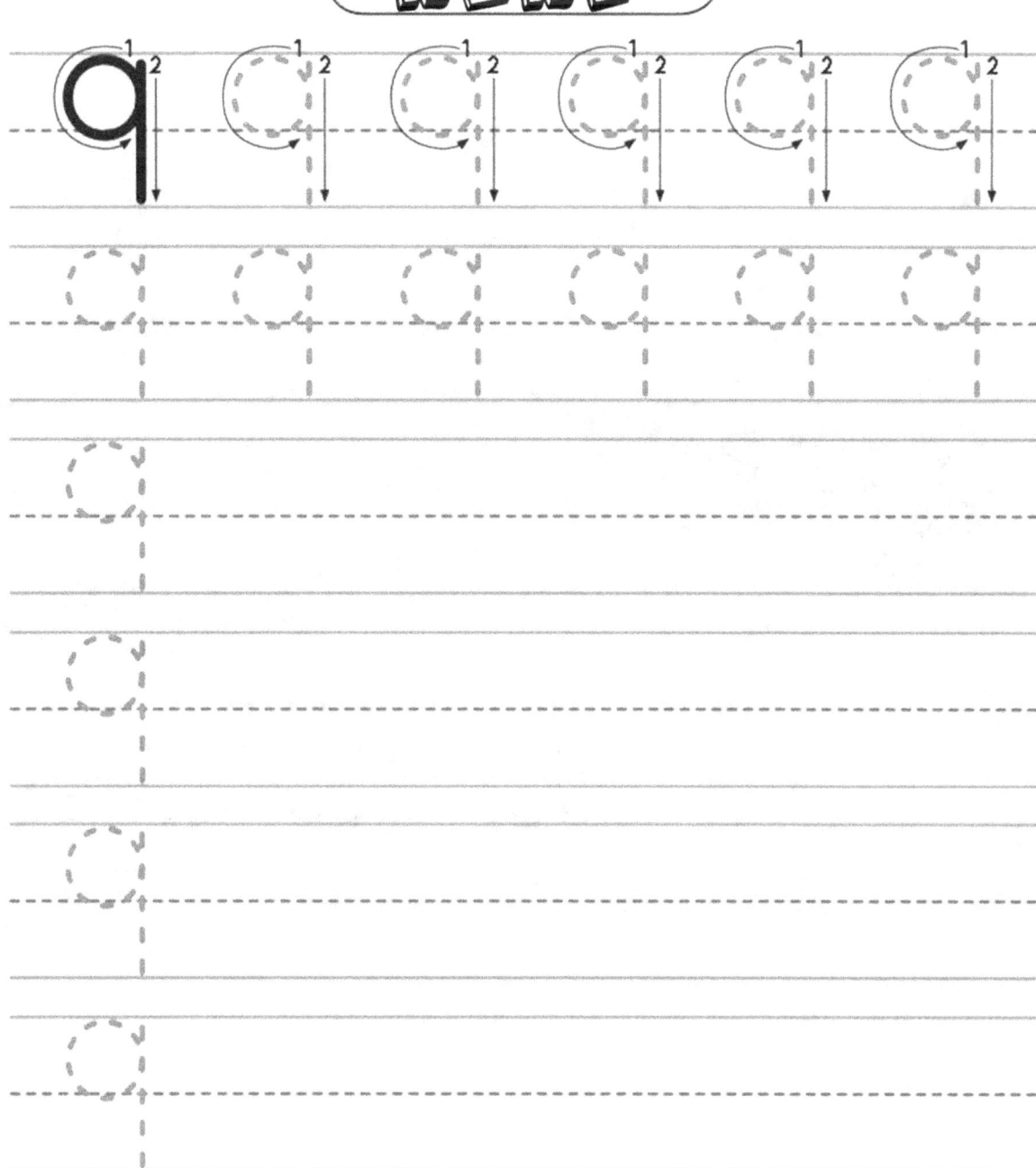

COUNT AND MATCH

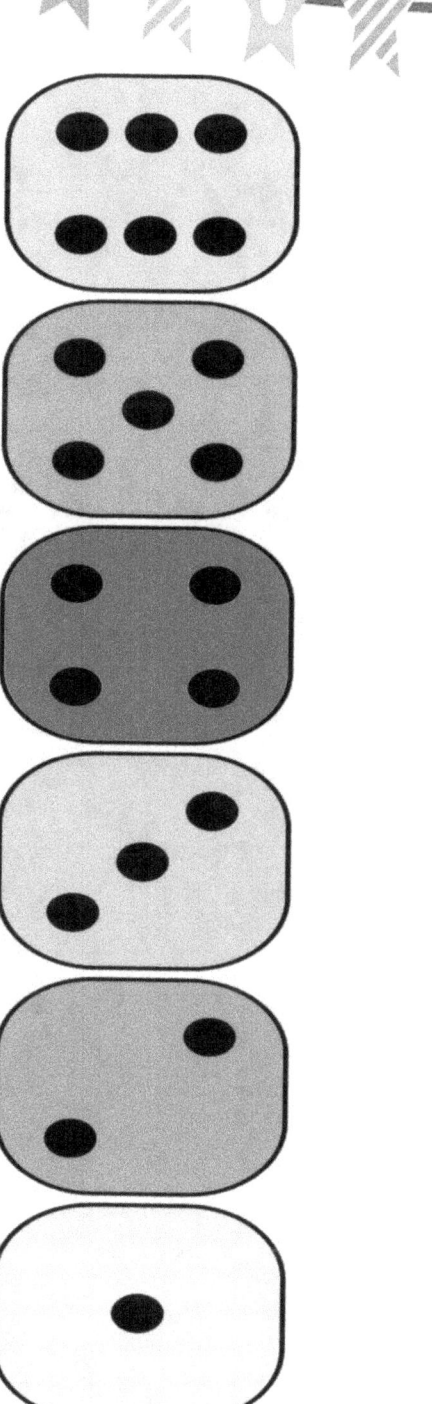

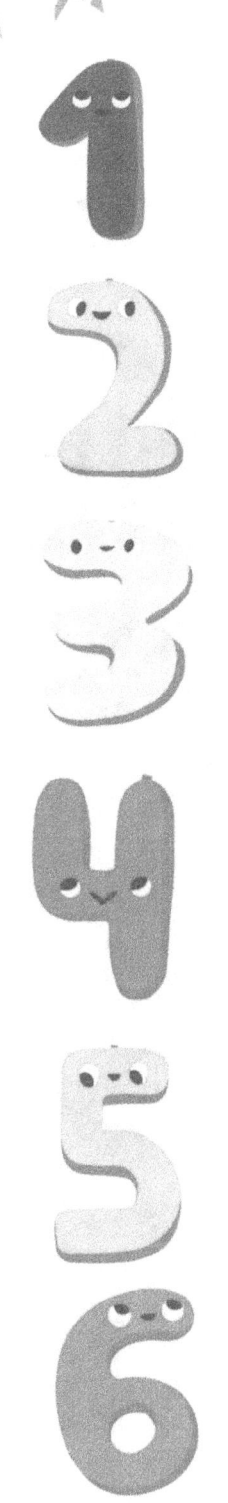

COUNT AND MATCH

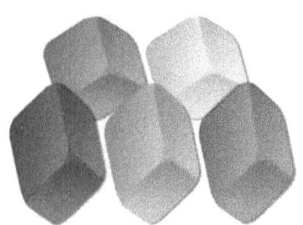

count the candle on each cake
Match with the correct number

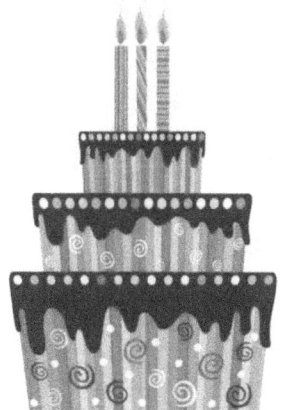

count the balloon   Match with the correct number

draw a line from the numbers to its'name

draw a line from the numbers to its'name

draw a line from the numbers to its'name

draw a line from the numbers to its'name

count the car   Write the correct number

0 1 2 3 4 5 6 7 8 9 10

count the doll   Write the correct number

count the egg

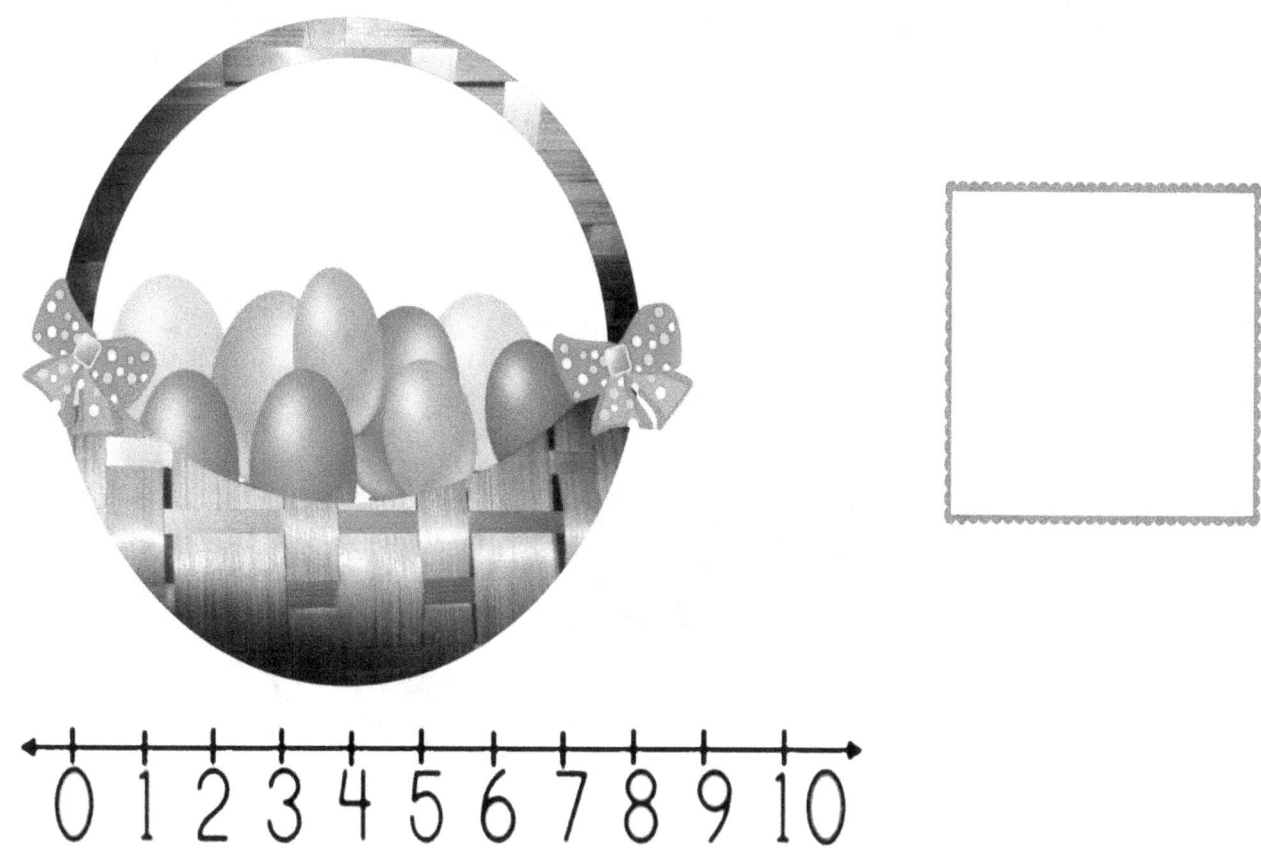

0 1 2 3 4 5 6 7 8 9 10

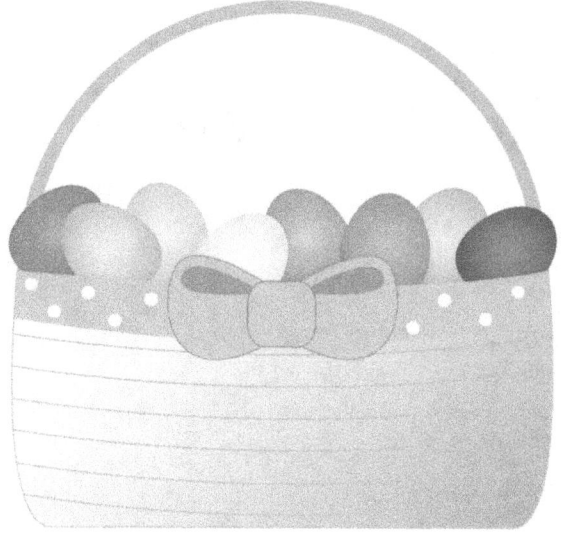

ANSWER KEY

COUNT AND MATCH

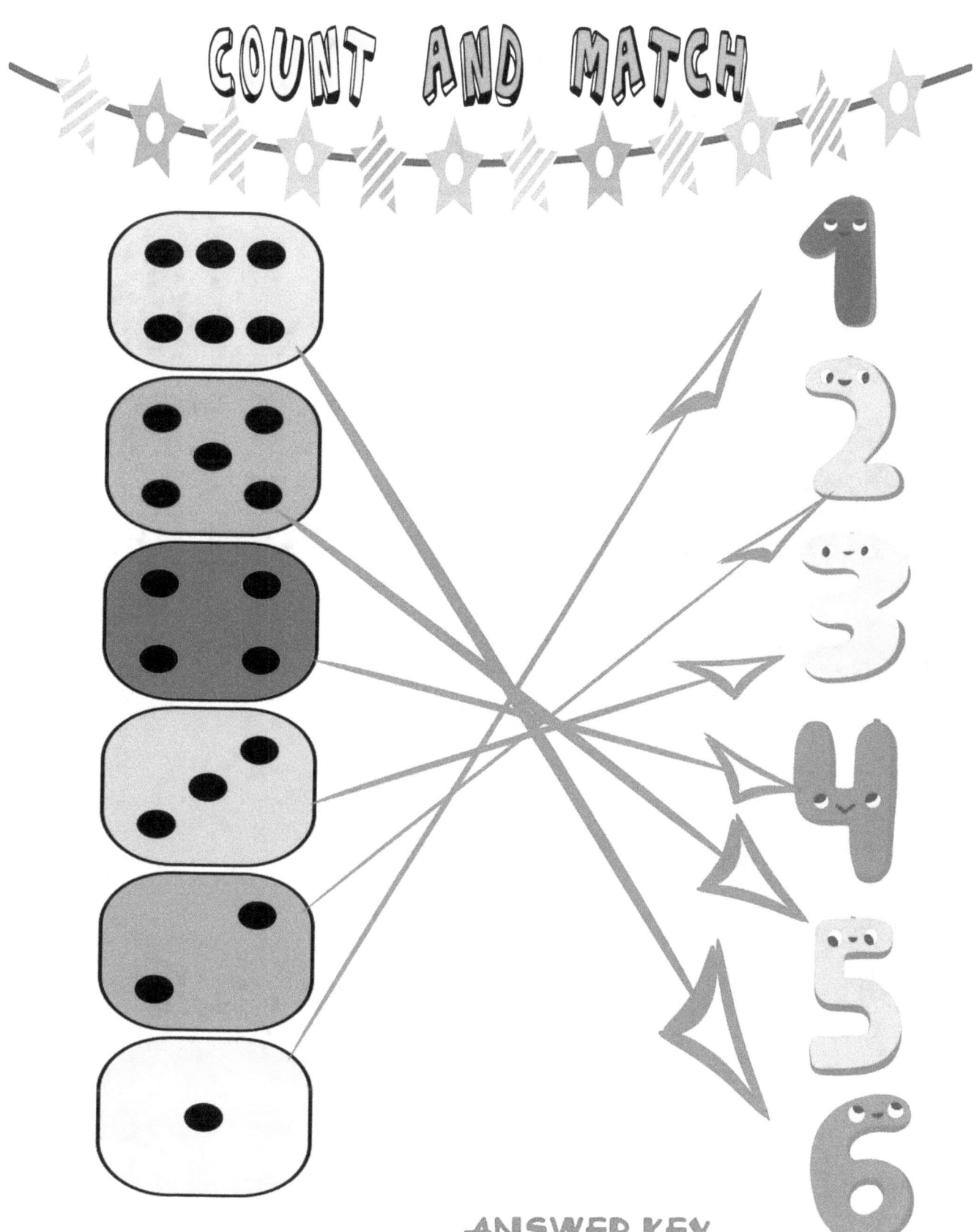

ANSWER KEY

COUNT AND MATCH

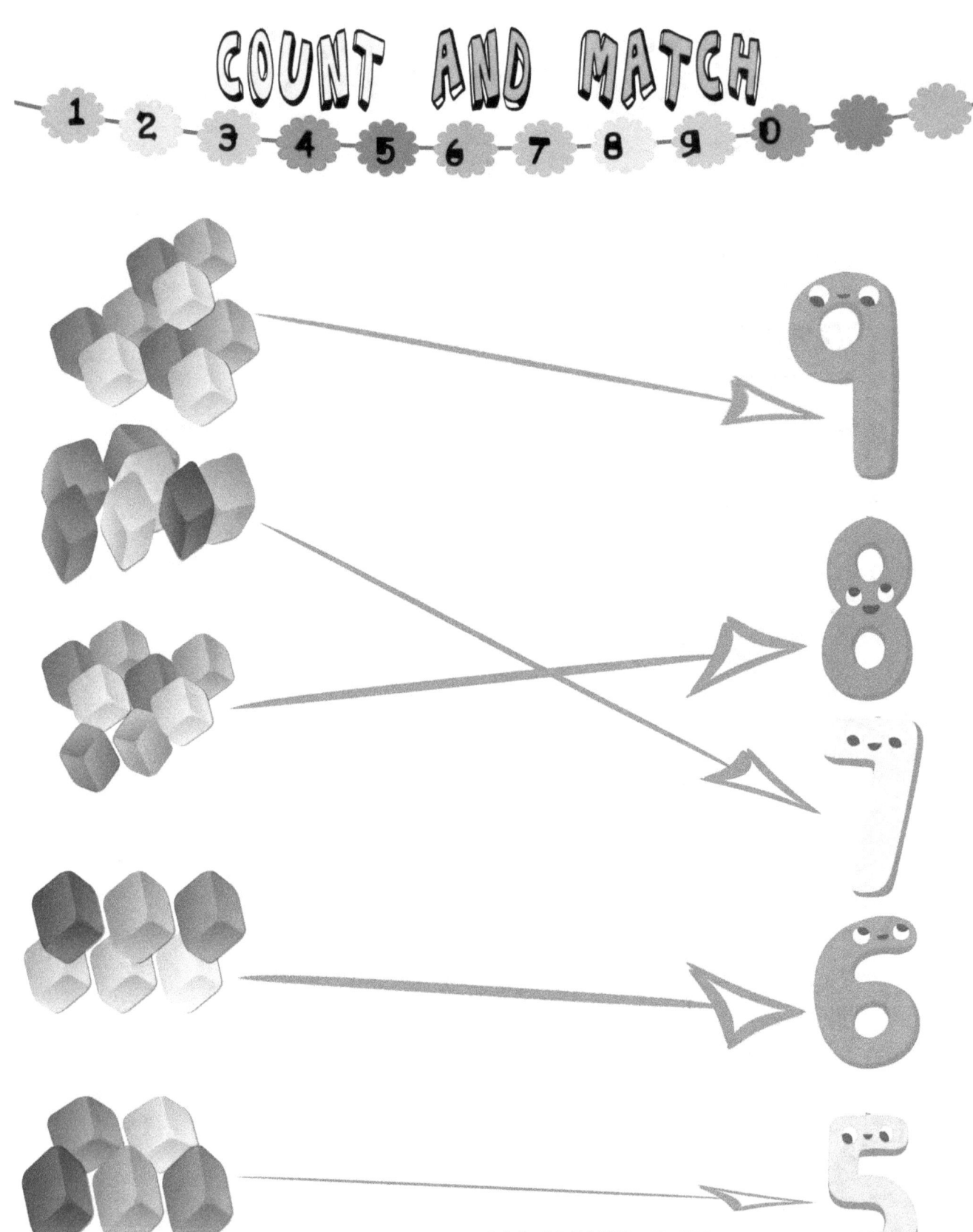

ANSWER KEY

count the candle on each cake
Match with the correct number

ANSWER KEY

count the balloon **Match with the correct number**

Answer Key

draw a line from the numbers to its'name

draw a line from the numbers to its'name
ANSWER KEY

draw a line from the numbers to its'name
ANSWER KEY

draw a line from the numbers to its'name

ANSWER KEY

count the car   Write the correct number

0 1 2 3 4 5 6 7 8 9 10

ANSWER KEY

count the doll   Write the correct number

0 1 2 3 4 5 6 7 8 9 10

ANSWER KEY

count the egg   Write the correct number

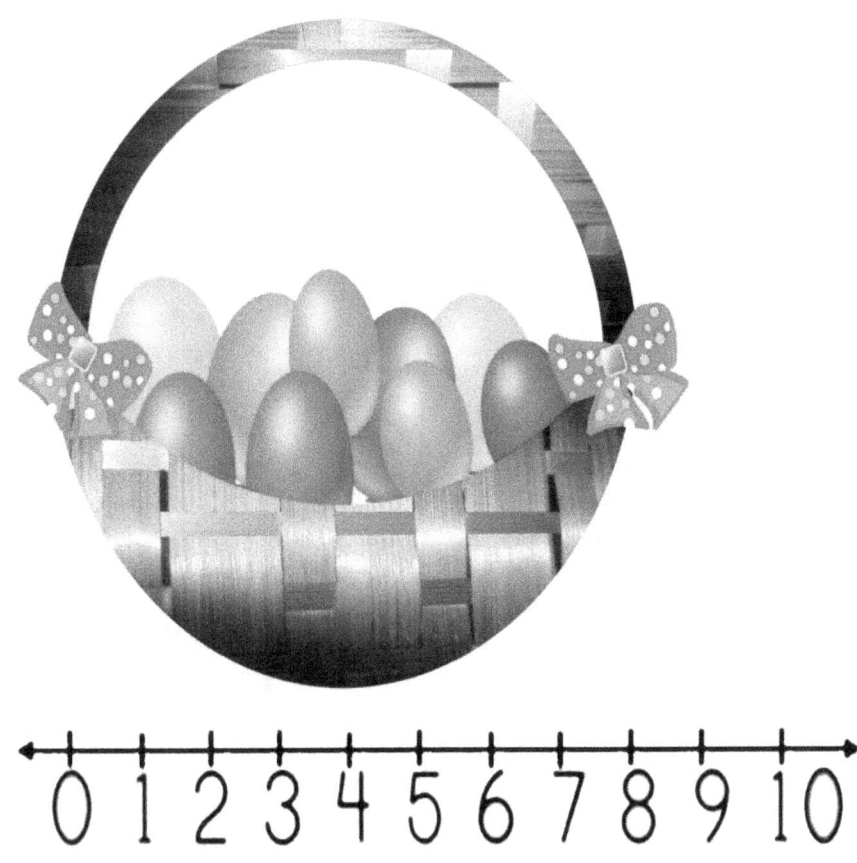

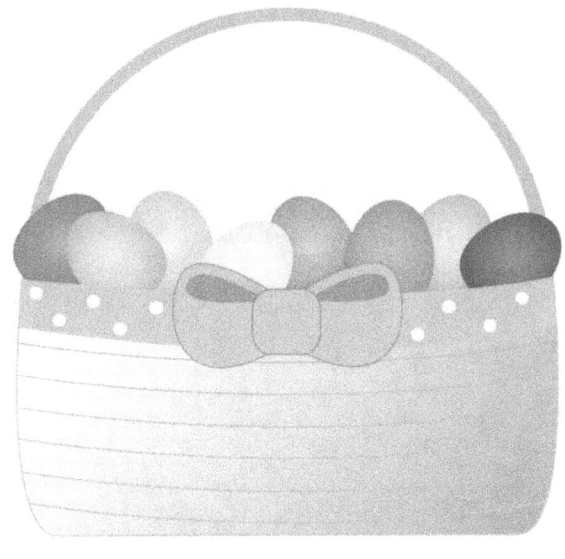

ANSWER KEY

www.ingramcontent.com/pod-product-compliance
Lightning Source LLC
Chambersburg PA
CBHW081614220526
45468CB00010B/2878

*9 7 8 1 7 2 2 9 6 4 3 3 7 *